Chewing the Cud

HENRY BREWIS

To my cousin $andy
From his
Cousin
June
✝

FARMING
PRESS

First published 1990
Reprinted 1990, 1991, 1993, 1996

British Library Cataloguing in Publication Data
Brewis, Henry
    Chewing the cud
    1. English humorous cartoons
    I. Title
    741.5942

ISBN 0-85236-211-0

Published by:
Farming Press
Miller Freeman Professional Ltd
Wharfedale Road, Ipswich IP1 4LG, United Kingdom.

North American distributor:
Diamond Farm Enterprises, Box 537,
Alexandria Bay, NY 13607

Typeset by Galleon Photosetting, Ipswich
Printed and bound in Great Britain by
Redwood Press Limited
Melksham, Wiltshire

Introduction

If you've quickly thumbed through this book already, deciding whether or not to buy it for your granny, mother, dad, or even yourself, – then you'll have realised it's a collection of poems and cartoons, reflecting an inside view of farming that even the environmentally conscious townie may not have seen yet.

The author has drawn his material from what often feels like several lifetimes spent on the brink of involuntary liquidation and mental instability, while in the company of fellow peasants (most of them similarly afflicted) and the assorted live and dead stock that conspire to make a career in agriculture a much more exciting way than most to end up broke.

Of course the poems do not seek to replace Wordsworth as required reading for English literature. They are just humorous verses that can sometimes rhyme a tale more eloquently than a load of long-winded prose.

Nor do the cartoons pretend to be works of art, only exaggerated pictures of exaggerated life down on the farm, – where 'the best-laid plans often end in a cock-up' (as Robbie Burns meant to say).

So if this book has any purpose at all, it's to reassure the reader that the daily disasters that befall him (or her) befall all of us who plodge about in wellies. We have to see the funny side now and then, or we'll go crackers. . . . You agree, – well that's a good start . . . read on.

Hartburn, 1990   HENRY BREWIS

Rhyme and Reason

I can't write clever verses like the modern poets do
that seldom rhyme and hide the reason as it tries to
    struggle through
perhaps I'm blind perhaps they play a game I wouldn't
    dare
but what of Edward Lear and Rupert Brooke and
    Walter de la Mare?

their's were simple poems learned unwillingly at school
pussy cats in pea-green boats, rhymed gently to a
    rhyming rule
'slowly silently now the moon, breathless on the windy
    hill'
metres meanings memories that linger gladly still

there was a certain music then, it danced upon the
    page
with steps we knew by heart and trip so lightly yet in
    dim old age
now some of us at least may hardly have the notion or
    the time
to go in search of reason where there isn't any rhyme

go back go back go further back and sit at mother's knee
when Mary had a little lamb and Bobby Shaftoe went
    to sea
when black sheep gave their wool away and Jack ate
    Christmas pie
. . . fresh flowers kept since childhood days
             . . . is rhyme the reason why . . .?

'. . . do y' think we should ask them t' come in, Willie, – it seems cruel t' leave them out there on a day like this . . .'

Lament of a Farmer's wife

Why don't you speak in the morning
y' miserable crotchety bloke
this can't be the way to start off the day
– are y' frightened you'd choke if you spoke . . .

you're not worth a damn in the morning
occasional grunt or a cough
and we sit there waitin' while you eat your bacon
– we might as well just wander off . . .

'don't expect y' to sing in the morning
or quote from the Latin or Greek
but perhaps we'd be stirred by the odd civil word
– is it too much to ask you to speak . . .

well you talk to the dog in the morning
and y' blather away to yourself
even a row with a bloody dead yow
– while we sit like a mug on the shelf . . .

you can talk on the phone in the morning
you always sound cheerful enough
but y' sit like a ghost when you're munching your
        toast
– in some deep agricultural huff . . .

do all farmers say nowt in the morning
is it part of the way they're brought up
there's about as much chance of a touch of romance
– from a knackered auld cross Suffolk tup . . .

can you never be bright in the morning
were you always this way as a lad
give a nod or a wink as I stand at the sink
– so the kids'll still know your their Dad . . .

we're not asking for much in the morning
when the news and the forecast are bleak
but with twenty odd years of blood sweat and tears
– SURELY T' GOD Y' CAN SPEAK!!!

6

'. . . it must be nearly spring . . . Dad's stopped talkin' . . .'

'. . . and you still reckon you've got nothing to declare, Sir . . .?'

'. . . isn't that just typical, – we get a few shares in Northumbrian Water, and they freeze the bloody assets . . .!'

Economical Truth

Have y' finished the lambin' young Willie
did y' have a reasonable do
ours went slow we've still twenty t' go
and aye we might have a geld one or two

    – Yes the whole lot were lambed in a fortnight
I was runnin' about on m' knees
they just wouldn't stop what a hell of a crop
and there's far too many damned threes

God that must be a record young Willie
did y' have nea deaths bonnie lad
nea staggers aboot nea slavery mooth
'cos I'll tell y' for nowt we certainly had

    – No we never lost nothin' that's honest
not even a back-body out
countin' the pets it's two hundred per cent
and oceans o' milk there is floatin' about

Well y' bugger that's magic young Willie
were your yowes tremendously fit
we seem t'end up with an arthritic tup
and Mules wi' three legs nea teeth and one tit

– Oh we always feed right through the winter
and of course the flock's beautifully bred
now just let me see was it seventy-three
when we last had a lamb that was dead

Well that's bloody remarkable Willie
but I think you're a lyin' auld sod
and I just came to say on this lovely spring day
that the postman's run over your dog. . . .

'. . . remember all that fuss about overworked doctors falling asleep during delicate operations . . .?'

'. . . what do y' mean, this is what happens when y' change to organic farmin', – you always *were* a filthy mess . . .!'

'. . . had a good day, Dear . . .?'

Out to Lunch

Well it's surely not the Hilton
or the posh Savoy Hotel
and you won't find swingin' city folk
'mongst the special clientèle

the wine list's rather limited
the menu's on the wall
but you're lucky every Wednesday
to get a seat at all

the dress is quite informal
like the intellectual chat
but you're viewed with grave suspicion
without wellies and a cap

it's the centre of the universe
where the dealer bids the trade
the auctioneer keeps singin'
and the grader makes the grade

peasants congregate in pickups
compare disaster and delight
'I'll get m' dinner at the mart pet'
moths attracted to the light

so what's the fascination
Egon Ronay's not to blame
the rain blows through the doorway
and the menu stays the same

   – well half the world will never know
   the other half have been
   and they've all got their places booked
   at Scots Gap mart canteen. . . .

'. . . he's the only buyer who managed to get here today . . .'

Dying Race

The yow is a creature I've studied too long
and still found no reason to burst into song
the beast is not witty, amusing well read
and gives the impression she'd rather be dead . . .

whether it's post- or just ante-natal
every disease will be more or less fatal
for the length of her life she will always conspire
to baffle the shepherd and quickly expire . . .

you may think the animal's quiet and tame
but nothing runs far when it's constantly lame
dose her inject her take care of her feet
but if science can cure her the bugger won't eat . . .

the fit ones have singles the worst produce three
all with a brain the size of a pea
and mother can't count her arithmetic stinks
so she may have had more than the stupid bitch
    thinks . . .

they're awkward they're brainless you must come to
    terms
that bugs and bacteria and most of all worms
will assist every sheep to escape from this earth
as quickly as possible soon after birth . . .

the breed doesn't matter they're all on a par
the trouble comes with them you needn't look far
their only appeal comes much later of course
with small new potatoes and tasty mint sauce. . . .

'. . . tell them they can start lambin' now, Dear, – I'm ready for them . . .'

Curried Egg

Take a woman who couldn't shut up
blathered on like a silly old hen
and ruined her dream her feminine scheme
to go roosting in number ten . . .

She may yet become Lady Edwina
she might still win a prize in the race
but the main one is blown and she'll always be
   known
as the woman with egg on her face. . . .

Take a pair of sparkling eyes
of a parliamentary rose
from the affluent south with a bloody big mouth
that didn't know when to stay closed . . .

Take a well-bred political lady
neat figure a nice shapely leg
who knew all about diet but couldn't keep quiet
when it came to boiling an egg . . .

DANGER

'. . . if that Currie woman sees this'n, Willie, – the whole mutton trade'll be knackered as well . . .'

'. . . nobody ever gets past *them* carryin' a bag, Sergeant . . .!'

'. . . well the question is, Willie, – do we move 'em out or muck 'em out . . .?'

Paradise Postponed

There's a heaven way up yonder
beyond the clear blue sky
where weary knackered peasants go
when it's time for them to die . . .

    where lambin' is a doddle
    and only twins are born
    where grass grows as you need it
    but you never cut the lawn . . .

where phantom jets don't practise
on quiet summer days
and you only have to wuffle once
to make quite perfect hay . . .

    where graders are redundant
    and you always 'top' the mart
    where no one's finished harvest
    before you get a start . . .

where there's never any pestilence
no sweaty feet or flu
and every time you're desperate
there's no one in the loo . . .

    where poor men get the pleasure
    and rich men take the blame
    no rates no tax no VAT
    and yowes are never lame . . .

where tractors don't get punctures
and muck just turns to gold
where in-laws never come to stay
and dogs do what they're told . . .

    where phones don't ring at dinner time
    and overdrafts are free
    . . . and if you swallow that my son
    you're crazier than me . . .!

'. . . gorn t' earth . . .!'

Wotsisname

I know him quite well I have done for years
it's just slipped through the hole in m' brain
what a ruddy disgrace I remember the face
but I've simply forgotten the bloody man's name . . .

   it's hopeless these days am I just gettin' old
   or perhaps it was always the same
   I've got no real excuse it's just no flamin' use
   no I cannot remember the gentleman's name . . .

went to school with the bloke a right canny lad
is it Jeffrey or Joseph or James
on the tip of m' tongue but it's not gonna come
I'm just hopeless rememberin' ordin'ry names . . .

   he farms up in the north of the county
   and gets four tons o' wheat so he claims
   I've been many a time yes I know the man fine
   but I can't get the damn man's ridiculous name . . .

now he married a lass from Newcastle
(this is drivin' me slowly insane)
and I think she's called Ruth but t' tell y' the truth
m' mind's gone a blank when it comes t' **his**
     name . . .

so I can't introduce y' I'm sorry 'bout that
I'm embarrassed perhaps y' can tell
I'll sneak off t' the loo and just leave it t' you
– 'cos believe it or not I forgot yours as well. . . .

'. . . we think he's had a Valentine . . .'

The Other Bloke

I met a man a well-dressed man
who drove a desk all day
computer phone and fax machine
overloaded pending tray
and civil service pension scheme
  I met a man financial man
  who did his sums all day
  bought and sold in Tokyo
  futures options come what may
  and watched his city fortune grow
I met a man a talking man
who poured out words all day
fact or fiction careless charm
kept his creditors at bay
and never seemed to take much harm
  I met a man an honest man
  who wheeled and dealed all day
  papered over business cracks
  made a million slipped away
  and never paid his income tax
I met a man a farming man
who still complained all day
blamed the weather and his wife
wondered why the rest would say
– we envy you your simple life. . . .

'. . . so which one of you terrorists is claiming responsibility for this outrage, then . . .?'

Caught

I crept in to watch the Test Match
put the kettle on t' boil
well I wasn't very busy
there was nothin' gonna spoil
and no one there to see me
not a soul would ever know
settled down before the telly
Vivian Richards in full flow
just three more for his hundred
we'd have to get him soon
what excitement there at Headingley
on a Tuesday afternoon . . .

then this sneaky little worm-drench rep
came knockin' at the door
even peeked in through the window
saw me sittin' on the floor
there was no escape he'd seen me
there was nothing I could do
a farmer caught red-handed
in the house at half-past two
oh the cruel humiliation
I was doomed to public shame
certified a waster
life could never be the same . . .

there was only one solution
and he put up quite a fight
but I buried him at tea time
when they came off for bad light. . . .

'. . . the first real sunbathing day, – and what do we get? . . . a posse of worm-drench reps . . .!'

May Morning

Oh what a beautiful morning
I feel like singing a song
if I could I would whistle jump over a thistle
today there's just nowt can go wrong . . .

all the spuggies and chaffs full o' chatter
a bright silver dew on the lawn
what a month is this May what a day is this day
it feels such a joy to be born . . .

cup o' tea and we're off round the stock then
breathe in the fresh morning haze
t' tell you the truth it's like inhaling youth
yes it's gonna be one of those days . . .

look the swallows are gobblin' up midgies
and the lambs chasin' up the dyke back
it's just quarter t' seven and God's in his heaven
and the townies are still in the sack . . .

even old Sweep's lookin' happy
such a fabulous day lies ahead
what's he found over there no the bitch wouldn't dare
— yes she would there's a yow lyin' dead. . . .

'. . . go on, Sweep, fetch 'er back, – pretend you're a bleedin' rottweiler . . .!'

'. . . and just think, if you were a yuppie off to the city in his Porsche, instead of an old peasant staggering over to the lambing shed in his wellies, – all you'd need would be a brolly and a filofax . . .'

'. . . oh don't rush away, lads, – I'm quite prepared to discuss the matter . . .'

Grave Concern

If our old grandad could see us all now
the way most of us seem to behave
he'd cough and he'd splutter, he'd swear and he'd
    mutter
and turn over three times in his grave . . .

  he would open the coffin look over the top
  and he'd surely say something uncouth
  'cos he generally said what came into his head
  and 'twas often quite near to the truth . . .

'' . . . the livin' is easy y' don't know you're born
y' should kneel and give thanks to your maker
if in doubt y' just spray or inject twice a day
and the barley's four tons t' the acre . . .

  if only your granny could see y' all now
  when she possed and polished and mangled
  when she hung out the weshin' fed twelve at the
      threshin'
  and the 'lectric was nowt but new-fangled . . .

mechanical this computerised that
y' don't know what hard work's about
now lo and behold the plan so I'm told
is to pay y' for just growin' nowt . . .

it doesn't make sense it doesn't add up
you'll surely grow lazy and fat
while y' sit on your bums and twiddle your thumbs
proper peasants weren't meant to do that . . .

who wants to be farmin' in this day and age
'wouldn't swop y' for five million quid
I don't give a damn I'll just stay where I am''
. . . and with that he would lower the lid. . . .

'. . . Peter, – you'll have t' do something about this hole in the ozone, – we've lost another one . . .!'

Where Have All the Seasons Gone?

Where have all the seasons gone
was it something that we said
that makes spring so reluctant now
to creep from winter's bed . . .
   that blows away the summer
   in one violent windy night
   and gentle mellow fruitfulness
   is such a brief delight . . .
or is it just a clever game
that Mother Nature plays
while painting pretty pictures
of all our yesterdays . . .
   warm images of boyhood
   with heroes ten feet tall
   making hay with Father
   when it never rained at all . . .
sweet adolescent daydreams
so much better than the truth
or the wayward wishful thinking
of a technicolour youth . . .
   is it just imagination
   so conveniently remembered
   all the bad bits best forgotten
   as we struggle through November . . .

all those working days and holy days
in rain and sleet and frost
and that snowstorm on Good Friday
when all the lambs were lost . . .
   but the rooks still go a-nestin'
   and the spuggie has his fling
   and even if the snowdrop's late
   there'll always be a spring. . . .

'. . . it's on mornings like this I realise I should've persevered with nuclear physics, Willie . . .'

Cash Flow

I've bought eighty-five store cattle
and a hundred yowes and twins
spread twenty tons of nitrogen
and now for all my sins
the bank rings up this morning
and with very little charm
says the overdraft is causing them
a certain mild alarm
it's the same old bloody story
you speculate or die
and the cash flow's one-way traffic
every year through to July
but we've started selling bullocks
sent some lambs up to the mart
we'll have barley in a week or two
when harvest gets a start
so the bank will have to wait that's all
let's face it they can't scoff
they loaned far more to Mexico
— and wrote the bugger off. . . .

'. . . you're wastin' your time lookin' for a rent rise this year, Colonel . . .'

Dog on Trial

Sweep come bye come bye t' me
good dog good dog haway
noo dinna just hide a hint that tree
or I'm tellin' y' son there'll be hell t' pay . . .
   here bonny lad noo that's the style
   easy 'n' friendly 'n' calm
   he'll m'be come bye if I try t' smile
   if he doesn't creep aal the way back t' the farm . . .
come in t' me son gotta be mad
quietly slowly come here
there's nea need t' look s' painfully sad
w' knaa for sure you're far ower dear . . .
   canny wee dog scratch his lug
   d' y' fancy he's got much eye
   beginnin' to think it's me that's the mug
   but we've got 'im this far we'll give 'im a try . . .
git away bye Sweep git away wide
gan out y' hopeless case
he nearly went I think he tried
but the sod seems t' have a grin on this face . . .
   y' wouldn't be foolin' me would y' Sweep
   any dog can have a bit fun
   but I'm tellin' y' noo git roond them sheep
   or so help me I'm off back yem for the gun . . .

once more just once afore w' give up
y' rotten brainless hoond
he's off he's away he's cowped the tup
have y' ivor heard such an aaful soond . . .
   shut up come bye and leave them be
   they'll be kebbin' all ower the hill
   Sweep sit doon come in t' me
   or I'll kill y' so help me I will I will . . .
where's there some string I'll hang y' now
come in t' heel aalright then stay
sit doon come bye and leave that yow
– oh hell I knew the bugger would run away . . .!

'. . . once he gets the rhythm goin', – he'll clip anythin' . . .'

Spittin' Image

Twenty-one and lovely
from a fashion magazine
with legs up to her armpits
and a figure like a dream
manicured and perfumed
the belle of every ball
hair of brightly burnished gold
and eyes that promised all
dressed to kill from morning
smiling lips of red
and what young men imagined
maybe better left unsaid . . .

   – then she fell for handsome Albert
a farmer big and bold
who said take off those clothes pet
and just do what you're told
throw off that fancy frock dear
you'll still look just as cute
in duffle coat and wellies
and this old boiler suit . . .

   – but twenty lambings later
and the figure that was there
is cunningly concealed now
'neath thermal underwear
there's a pet lamb in the kitchen
and the forecast is for sleet
there's a delicate aroma
of wet dogs and sweaty feet
now she's forty-two and cuddly
and her hero's less enticing
the gilt is off the gingerbread
the cake has lost some icing . . .

   – then enter lovely daughter
the belle of every ball
with legs up to her armpits
and eyes that promise all
manicured and perfumed
from a fashion magazine
. . . with the nearest thing to Albert
that her mother's ever seen. . . .

'. . . alright, clever clogs, – *you* try pullin' wild oats in a howlin' gale . . .!'

'. . . I was a farmer once y' know, – but I don't owe anybody anything now . . .'

'. . . this is it, Son, – this is the alternative land use we've been lookin' for . . .'

Poor Thing

How cruel the killing of that beast the lady wrote
shot through the head the knife drawn swiftly 'cross
    the throat
poor thing deceived betrayed to satisfy the thief
who took a life for sirloin fillet brisket beef
and even stole the skin to make a winter coat . . .

the lady wept her righteous tears and bowed her head
and prayed for creatures great and small condemned or
    dead
then without a backward glance into her tortured soul
she fed her cats their lamb and chicken liver casserole
  and thus enriched she took her cup of Bovril up to
    bed. . . .

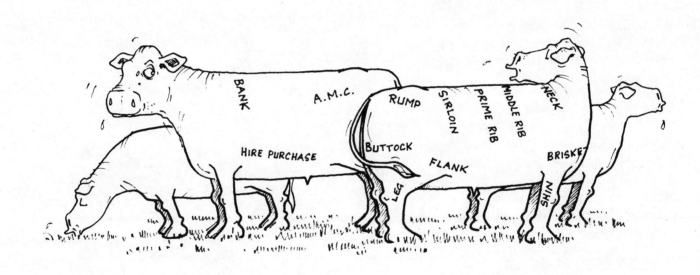

'. . . obviously we've been to different conferences . . .'

Bath Time

What we need is a modern new dipper
with a well-designed tubular run
where the sheep just dive in
have a quick little swim
and stroll out up the steps when they're done . . .
   there's no need to wear waterproof y-fronts
   and y' don't get the stuff in your eyes
   the old sheep don't suspect
   what's coming up next
   and their bath is a total surprise . . .
not a bit like our ancient old bath tub
where the farmer gets dipped with the sheep
as he stands in the hole
like a soggy dead mole
and the tank develops a leak . . .
   then the water's reduced to a trickle
   the yowes go through just when they please
   you might blast and you'll damn
   'cos you can't cowp the ram
   and you're frozen to way past your knees . . .
what we need is a new modern sheep bath
with plenty of dip water in it
but I have t' say now
we don't have a mule yow
who'd survive for the full legal minute. . . .

'. . . hey, – where's that bloke with the stop watch gone . . .?'

Set Aside

Eighty quid an acre isn't any good to Tom
he grows four tons t' the acre and he's going like a
    bomb
he's got two big fancy combines and a fleet of bright
    machines
set-aside's for someone else who paints a different scene
    — meanwhile the bank is helpful but our man knows
    very well
    to simply stand where he stands now he'll have to
    run like hell

Eighty quid an acre won't impress a bloke like Dick
he grows three tons t' the acre and his system does the
    trick
he feeds a hundred cattle and a thousand Suffolk lambs
and the cereals are just a part of one great master plan
    he's never made a fortune and perhaps he never will
    but set-aside would hardly pay the annual int'rest
    bill

Eighty quid an acre doesn't seem a lot to Harry
but at two tons if he's lucky and some knackered gear
    to carry
the plot becomes attractive especially when the figures
show he's slowly going nowhere as the overdraft gets
    bigger
    when all you need is twenty fags and half a pint of
    beer
    eighty quid for growin' wickens isn't such a daft
    idea. . . .

'. . . but it's just a simple balance of payments problem, Mr Thompson, – what the Chancellor might call a temporary blip . . .'

Telephone Gossip

Hallo there Florrie it's just me
felt like a little chat
no I haven't any startlin' news
no scandal nowt like that . . .
I thought I'd just see how y' were
not seen y' for a week
yes Charlie's fine well more or less
of course he hardly ever speaks . . .
that' right your Willie's not much better
there's not one t' beat another
m' father was the ruddy same
accordin' to m' mother . . .
'course the trouble's always money
y' would think we must be broke
just ask him for an extra quid
the bugger has a stroke . . .
last week he went berserk y' know
you'll ruin me he said
went on until the epilogue
and wouldn't come to bed . . .
he was sulkin' still at breakfast time
complainin' moanin' on
electric water rates and rent
and where's it comin' from . . .
I don't know how I stand it
work m' fingers t' the bone
he goes stark ravin' mad each time
he hears me on the phone . . .

'served 'im right last Sunda'
he's been huffy ever since
instead of beef and yorkshire pud
I gave 'im lumpy mince . . .
'told 'im straight it wasn't on
the money's just no use
he drinks as much as he gives me
and I just get abuse . . .
he's down the pub again of course
with all his boozy mates
'came in half past twelve last night
in such an awful state . . .
oh God I'll have t' go I hear 'im now
ta ta then see y' Monda'
    — what's that pet no just picked it up
    some woman dialled wrong number. . . .

'. . . think nothing of it, Mother, – he always gets a little irritable at harvest time . . . and haytime . . . and lambing time . . . and . . .'

All to Pot

You will seldom meet a farmer
who is into marijuana
and similarly few
are accused of sniffing glue
I suspect you'd say the same
on the subject of cocaine
and it never was our scene
to indulge amphetamine
of course you may have seen us
injecting livestock intravenous
giving patent stimuli
to a sheep that wants to die
there may be those who take a trip
on a drop of winter dip
or a psychedelic ride
on a whiff of fungicide
and some may get a kick
from a tasty min'ral lick
but unlike the hippie masses
we've better uses for our grasses
we don't need the pills and pot
to improve our rural lot
all we need for life sublime
is some luck at lambing time. . . .

'. . . I'm tellin' you guys, – the things that grow in set aside, – unbelievable, – way out,
man . . .!'

Dream Holiday

She tried her sweet persuasion
every single year come June
but it never seemed convenient
much too late or just too soon . . .
   there was always some disaster
   that he couldn't leave alone
   a yow a cow some problem
   that needed him at home . . .
just a weekend in Majorca
she would plead right from the heart
a kind of second honeymoon
this time without a mart . . .
   we could lie there in the sunshine
   it might cure your chesty cough
   we'll drink wine and dance fandangos
   but by then he'd nodded off . . .
well townies always go abroad
and he stirred enough to say
those buggers all have pensions
and holidays with pay . . .
   she would try again next morning
   like talking to a log
   who would spray the mildew
   who would feed the dog . . .

there's that heifer with mastitis
and half the lambs are lame
I've never needed holidays
and Father was the same . . .
   she pestered and she pleaded
   more in anger than in sorrow
   till he smiled and said I'm sorry pet
   – the harvest starts tomorrow. . . .

'. . . George, – just don't ask . . .'

'. . . complete breakdown, I'm afraid, – believes the world is controlled by something called a Mule yow . . .'

'. . . you may not believe this, Ferdy, – but I once actually *met* a cow, y' know . . .'

One more Gone

There once was a farm on a hill-top high
where for hundreds of years as the seasons rolled by
the pigs and the hens and the blue suckler cows
the barley the wheat and the flock of mule yows
were taken for granted and no one asked why . . .

there were hard-working farmers with hard-working
   women
bringing up families with what they were given
through sunshine and showers snow blocking the lane
going bust at the bank and starting again
on the two hundred acres that gave them a livin' . . .
  'til out of the mist in the Thatcherite era
  with the folks from the town getting nearer and
    nearer
  when everyone else talked of nothing but money
  when young yuppie bees buzzed in hives full of
    honey
  the view from the hill became suddenly clearer . . .
it was then that the derelict pig sty and barn
the stable the hemmel took on a new charm
with planning permission who wants to grow more
there's an upwardly mobile young man at the door
who can't wait to develop this family farm . . .
  now there on the hill for many a year
  there won't be a peasant and who sheds a tear
  there are several accountants a goat and a horse
  and some smart BMWs parked there of course
  — and we've just seen another old farm
    disappear. . . .

'. . . bugger off you lot, – I'm sellin' your pad to a stockbroker . . .!'

Life Cycle

you might think that's the end of the story of course
(if you've never had sheep of your own)
but through sunshine and snow there's a long way to
    go
'til that lamb is anywhere near fully grown . . .

the nuisance will hardly be out of the pens
he just has t' be dosed every week
then his mother is clipped and the whole lot are
    dipped
and we haven't begun to consider their feet . . .

Born to an ageing arthritic mule
on a filthy Wednesday night
lay by the fire then out in the byre
'til we figured his mother could feed 'im alright . . .

we gave him colostrum injections and pills
every damn thing he might need
then we ringed the poor sod on his tail and his cod
and turned him away with the rest on the seeds . . .

the trauma goes on the whole summer long
'til the sheep simply graze in your head
and you know every morning without any warning
you could easily find two or three lyin' dead . . .

but with luck it survives till the end of July
some reward can at last be expected
so you're off to the mart with a song in your heart
– and the bugger's too fat it's rejected . . .!

'. . . don't worry, she'll take it eventually, – your Father knows all the old traditional
techniques . . .'

Yuppie Ballad

Young Nigel was a banker and Fiona was his spouse
charming and successful with a large suburban house
Mercedes and jacuzzi and a bath shaped like a heart
and an avocado bidet to wash their high-class private
　　parts . . .
　A big secluded garden with some rare exotic blooms
　a little man who cut the grass on Friday afternoons
　they knew the Jeffrey Archers they were Tories to the
　　bone
　and standing on the patio a Margaret Thatcher
　　gnome . . .
European holidays Paris Rome Madrid
the house bought only yesterday worth half a million
　quid
sophisticated people of that there was no doubt
even called each other 'darling' when no one was
　about . . .

　Exclusive private suppers with enlightened special
　　mates
　drinking château-bottled wine and nibbling After
　　Eights
　talking loud and laughing seldom listening to a word
　the sound of loads o' money was the only noise they
　　heard . . .
Call them yuppies call them townies call them
　anything you wish
the bold new generation and quite seriously rich
confident articulate that upward modern mixture
call them greedy if you want to but I'm sure you get
　the picture . . .
　Then one day young Nigel announced to all and
　　sundry
　they were moving from the city to a place out in the
　　country
　''it's the quality of life y' know the air is fresh and
　　clean
　we're going back to being peasants, we're going
　　absolutely green . . .
Oh no more mad commuting on the nine-to-five
　express
no more wheeling dealing in the rat race to success
we've bought a lovely farmhouse isn't even on the map
halfway between the old A1 and some place called Scots
　Gap . . .

You simply must come up and see us the
    countryside's quite charming
and it won't take very long my dears to master
    simple farming
we'll have to rearrange some things and get the
    system right
but one understands the natives aren't particularly
    bright . . .
I expect there's lots of subsidies and set-aside of course
and Fiona and the children say they'll have to get a
    horse
naturally we'll hunt a lot and shoot the wayward bird
so if you see our special friends be sure to pass the
    word. . . .''

There was no one quite so confident
there was no one quite so smart
as the slick city gent with more money than sense
as he stood there at Scots Gap mart . . .
but you'd be surprised how easy it is
to make a right mess of it all
and end up with a tup who can hardly stand up
with one inadequate ball . . .
you'd be surprised how simple it is
for an arrogant city-bred gent
to combine his corn on a fine dewy morn
at somewhere near forty per cent . . .
to purchase a pen of mule ewe lambs
straight off the Rothbury heather

and find out too late that four out of eight
are what's known in these parts as a wether . . .
you'd be surprised how easy it is
to follow your vet to the byre
just as that cow decides that right now
is a suitable time to expire . . .
to cut down some hay on a rare summer's day
and curse your astonishing luck
as it rains all July and you're left wonderin' why
you've got forty acres of muck . . .
disease in the wheat sheep's rotten feet
pestilence plague it's not funny
and before you could say have a nice day
poor Nigel had run out o' money . . .

Young Nigel was a banker and Fiona was his spouse
charming and successful with a large suburban house
but that of course was yesterday when Nigel was a toff
now he's sweeping up clarts at Scots Gap mart
and Fiona's buggered off . . .!

'. . . what's this then lads, – some sort of harvest festival is it . . .?'

The Drought of '89

You'll maybe not remember the great drought of '89
when it didn't rain from New Year's Day
right through t' Christmas time
not a single April shower
nor a foggy foggy dew
never wore m' wellies
and the sky was clear and blue . . .

you'll maybe not remember but the reservoirs ran dry
and the air became so heavy
that the birds refused t' fly
just a trickle in the rivers
fish were floatin' on their backs
the land so bare and crusty then
old yowes fell down the cracks . . .

you'll maybe not remember such a long and serious
        drought
when a raindrop on the window
brought a wild excited shout
when making hay was easy
and we got a field of corn
at under twelve per cent that year
and never cut the lawn . . .

you'll maybe not remember but perhaps you might've
        heard
that the weather man McCaskill
was almost lost for words
he put the blame on aerosols
but we didn't really care
as we sunbathed in the garden
'neath the crumbling ozone layer . . .

you'll maybe not remember but in the nick of time
someone had a vision
from the city came a sign
the world and we were saved again
from an unexpected quarter
St Margaret of Ten Downing Street
privatised the water. . . .

67

'. . . lovely we(a)ther, Charlie . . .'

Profile

He's the man the media looks for
the expert of the day
on every new committee
with quite a lot to say
he's never seen in wellies
or a spanner in his hand
'cos you'll seldom find him farming
at home upon the land
he's much better when he's talking
understands the EEC
the green pound holds no mysteries
for men as smart as he
pontificates on telly
the early breakfast show
answers awkward questions
that the rest of us should know
his voice more condescending
as his fame begins to rise
but back at home he's knackered now
surprise surprise surprise
*   — yet our man's not at all dismayed*
*   he'll never be redundant*
*   he'll set aside the bloody lot*
*   and be a farm consultant. . . .*

'. . . aye they're not bad sheep, Willie, – but y' can't be too careful these days . . .'

Chernobyl Fallout

I always knew they were trouble
right from the very start
lambin' or clippin' dosin' or dippin'
always determined t' break your heart . . .

I always knew they were stupid
most of 'em born that way
they even arrive more dead than alive
and often reluctant to stay . . .

I always knew they were awkward
they can die from a terminal sneeze
there isn't a vet who has mastered them yet
or heard of every disease . . .

I always knew they were wicked
nothin' the least bit attractive
and now the last straw it's much worse than before
half the buggers are radio-active . . .!

'. . . and y' reckon that's m' Dad, – y' have t' be jokin' Mother . . .'

Short Story

It's a sad little story
no hope and no glory
you may shed a sensitive tear
for one stroke of the knife
has determined my life
will be spent as a Charolais steer . . .

they crept up behind
with a cut so unkind
I was pruned from a proud family tree
I knew Mother quite well
but when Dad rang the bell
he couldn't stay round to meet me . . .

it's all too frustrating
this careless castrating
my future is brief and quite dull
I'm condemned to the grade
when convinced I'd have made
a perfectly marvellous bull. . . .

'. . . don't worry, it's alright children, – your Father never sleeps well after he's been clipping all day . . .'

Highland Island

There's a little highland island
to the west of Firth o' Clyde
where the Tighnabruich turtles sleep all day
and the Knockenkelly eagle takes the north Atlantic
    seagull
for a fancy feathered reel across the bay

  where the salty shiny briny
  to the south of Kyles o' Bute
  plays a melancholy liquid lullaby
  where the tiny tartan fishes blow the sailors sloppy
    kisses
  and the kilted Scottish salmon learn to fly

where the ocean has a notion
to caress Ardscalpsie point
and tumble through the long Kilbrannan sound
not a million miles from Wemyss lies the island of her
    dreams
and that's the place the lady can be found

on her private highland island
sits the mermaid of Inchmarnock
with her happy hairy consort by her side
and they look out to admire setting suns on old
    Kintyre
as the dolphins bring the kids home on the tide. . . .

'. . . y' don't think he'll take any notice after your language in the sheep pens this morning
do y' . . . ?'

Farm Secretary

No it's not much like working in Milton Keynes
for some multinational gnome
who smokes fat cigars drives a company car
and is only a voice on black cordless phone . . .
where the view from the window is concrete and glass
if indeed there's a window at all
just a green plastic plant and some fish in a tank
and a picture of Maggie stuck up on the wall . . .
so who wants to labour from nine through to five
in a prison so lacking in charm
when just down the lane there lies fortune and fame
on your friendly old neighbourhood countryside
    farm . . .

yes that secretarial farming career
could be the bright future for you
where the heart will beat faster and daily disaster
ensures that the language is colourful too . . .
so brush up your book-keeping typing and VAT
your accounting taxation and law
learn how to persuade that young rep he'll be paid
in a month if he takes his big foot out the door . . .
it's a joy it's a doddle it's money for jam
as long as you quickly can tell
the diff'rence between a wild oat and a bean
and whether the lambin' ain't going too well . . .

you can learn about clawbacks and quotas and costs
conservation computers and such
while a cow lying dead at the back of the shed
may be not the right time to be asking too much . . .
remember a farmer's reluctant to talk
he's more likely to grumble and groan
but treat the poor sod as if he was God
and he'll quickly assume that you're one of his
    own . . .

yes a far better world could be beckoning now
even though it's a strange way of life
but you'll soon get the knack and you'll never be
    sacked
if you always stay friends with the boss's dear
    wife . . .!

77

'. . . I'm hijackin' y', bonnie lad, – cut twenty-five acres for me, – then y' can go anywhere
y' like . . .!'

'. . . I suppose they must be the ones I promised t' pay as soon as we finished the harvest . . .'

'. . . so can I take it, Mr Thompson, that you don't quite share my optimism . . .?'

Billet Doux

Dear Mister Banker how do you do
find enclosed a nice big cheque a little overdue
the barley's all been paid for
you'll be quite pleased to hear
and the cash for the wheat
should be with you next week
so we might struggle through to next year . . .

Dear Mister Banker regards to you and yours
but before you get excited we'd like to buy some stores
we need another tractor
and some gimmers should be bought
so you'll see very clearly
I remain yours sincerely
in need of some further support . . .

Dear Mister Banker I trust you're feeling well
because I have to tell you sir we've nothing more to
     sell
at least not for a month or two
'til when I must depend
on you and the missus
so here's love 'n' kisses
from a truly devoted old friend. . . .

'. . . it's the complete tranquillity of the countryside I find so uplifting, Mabel . . .'

Wake Up

The countryside's awakening
stretching turning yawning
signs of early restlessness
this mellow spring March morning . . .
the light is creeping silently
across the dark green fields
no noise no interruption
as night time gently yields . . .
only wind through naked trees
as blackbirds tune their beaks
undisturbed becalmed the land
rises from its moon-kissed sleep . . .

then Boom and bloody Boom again
tranquillity is broken
sparrows crows and lazy dogs
are rudely now awoken
Boom the bangers strike Boom Boom
the rural overture begins
armageddon greets the day
pigeons check their next of kin
Boom Boom the 'scarers' thunder
ye gods they're everywhere
no wonder there's a gaping hole
in Mother Nature's outer layer
then suddenly the Phantoms come
screaming screeching in from hell
a million miles an hour
just a yard above the fell
Boom and ROAR and CRASH and BANG
angry shepherds stand and swear
at decibels that crack the dawn
— there'll be more peace in Leicester Square . . .!

'. . . aye, I suppose the townies will be out in their thousands on a day like this . . .'

Flotsam and Jetsam

There's a generation's flotsam
almost every farmyard's got some
just around the corner tucked away
in a rampant bed of nettles there's a pile of rusty metal
that cost a bloody fortune only yesterday . . .

   There's some crazy zigzag harrows
   and a wheel-less leg-less barrow
   and what used to be a gang of Cambridge rollers
   but all the rings are cracked and the bearings prop'ly
     knacked
   in the struggle for a tilth among the boulders . . .

There's a holey diesel tank
and a fork without a shank
half a dozen pails without a bottom
seven wheels without a tyre worn-out shovels from the
    byre
in the days he kept a cow almost forgotten . . .

   There's a plastic bag of strings
   and a bed frame with no springs
   and a well-corroded fertiliser spinner
   a tea pot with no handle wire netting in a tangle
   and an oven that once cooked a thousand
     dinners . . .

There's what's left of that first reaper
and it might've been much cheaper
if he'd never even bought that ruddy plough
if we added up the cost of the broken and the lost
well – he might've been a millionaire by now . . .

There's a generation's jetsam
surely everybody gets some
bits of history lying rusting night and day
some of yesterday's inventions and the very best
    intentions
of the farmer who was last to pass this way. . . .

'. . . have a good round, Sweetheart . . .?'

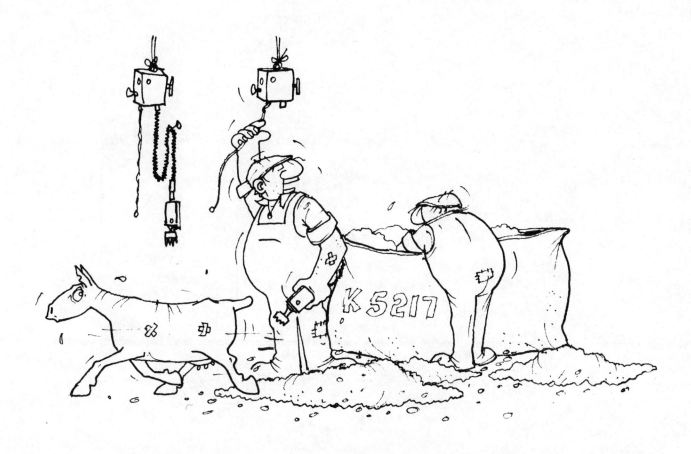

'. . . that's the last 'n, Willie, − hundred and eighty-seven fleeces, three lugs and a tit since breakfast time, − not bad eh . . .?'

'. . . and did you manage to finish the field, Dear . . .?'

Return of the Peasant's Daughter

She was only a poor peasant's daughter
from the land where the Charolais roam
but her Dad was uncouth and to tell you the truth
she just couldn't wait to leave home . . .
the language at lambing — disgraceful
the kitchen was just a disaster
all those wellies and clarts rotten silage and marts
she'd elope but nobody asked 'er . . .
when the weather was foul so was Daddy
when the combine conked out he went spare
she would work out which day to keep well out the
    way
and when to say nowt and beware . . .

    then the lass got her GCSE
    passed her test at the second attempt
    her Dad said "work here" she said "no bloody fear"
    and she patted the collie and went . . .
    she soon got a job in the city
    filofax brollie and suit
    she came home at weekends with her smart townie
      friends
    went for walks in her red kinky boots . . .

then she married a well-heeled accountant
upwardly mobile with class
she learned how to speak without parting her teeth
and a mouthful of crystallised glass . . .

they soon moved from the town to the country
an old farmhouse ten acres a shed
and you've guessed it of course they now have a horse
and six sheep but one of them's dead . . .
there's a hamster and seventeen pullets
three heifers who each have a name
— now she'll tell you with charm that having a farm
is a "frightfully int'resting game. . . ."

'. . . yes, I agree, – you *do* seem to be rather unlucky . . .'

Choral Morning Song

They all meet up in committee every morning
discussing what the day might hold in store
chattering they sing of a life spent on the wing
or perhaps who shared their nest the night before . . .

    I think sparrows all delight in idle gossip
    they just blather while there's daylight in the sky
    not listening to a word from any other bird
    who wants to stop and talk while passing by . . .

full of self-indulgent agitated banter
like an endless parish council AGM
through the length of every day every one must have
    his say
from the hawthorn hedges filled with cock and hen . . .

what a splendid little fella is the sparra
such a happy anxious chirpy busy bloke
as common as they are there's just nothing on a par
with this character among the feathered folk . . .

that bird's always there to greet me in fine fettle
his the only song I need for early warning
let the chorus be complete tune up all those other beaks
but give me spuggies in the garden every
    morning. . . .

'. . . I wouldn't worry too much, Sir, – we never anticipated the rural vote would be very significant . . .'

The Rt Hon. Member

He has a ready answer always
the question doesn't matter
he'll still come up with half an hour
of painful pompous patter . . .

no there never was a subject
he considered out of reach
and needed no encouragement
to make another speech . . .

he can blather on regardless
taxation crime defence
farming social services
'til none of it makes sense . . .

he's always on the telly
and he's quoted in the press
this selfless public servant
bravely sorting out the mess . . .

rearranging all the chaos
solving every little riddle
of looney left or heartless right
or indecisive middle . . .

only he can give us justice
only he can tell the truth
prosperity for everyone
a future for your youth . . .

only he'll defend the nation
any promise in a storm
more money for the NHS
to keep the old folk warm . . .

principles convenient
a condescending smile
deaf to contradiction
the parliamentary style . . .

so give the man a knighthood
a nation's just reward
he's bored the pants off all of us
now send him to the Lords. . . .

'. . . y' shouldn't have waited up, darlin' . . .'

Celebration

Does it really start at forty
has the livin' just begun
are there mountains still for climbin'
are there races yet to run
are there fortunes to be fashioned
will the world make better sense
will the grass be any greener
on the far side of the fence
or are troubles just beginning
will the wife demand too much
will his swing get even shorter
will he loose his putting touch
can he stand the mounting pressure
can he face the awful truth
can he still pot all the colours
like he did in long-lost youth
or will the years defeat him
is the magic sadly gone
just a quick disintegration
just a wreck at forty-one
will his gammy leg support him
is his back a problem now
can he catch a wayward gimmer
can he catch an eight-crop yow . . .?

– have a brandy and Sanatogen
smile bravely through the tears
buy a decent walking stick
change down to bottom gear
the second half is still to play
against the wind it's true
so blow out all those candles dear
we've still got things to do . . .!

'. . . the bitch wouldn't sit still, – so I had t' give her a local anaesthetic . . .'

'. . . all is safely set aside then is it . . .?'

'. . . he had a moderate lambin', a rotten silage crop, a bad harvest, slugs in his winter wheat, – and now she starts talkin' about Christmas shoppin', – it's just been too much for 'im . . .'

Fond Farewell

The kids nearly cried when the wagon arrived
and Charlie went off to the mart
well he was a bullock we'd fed from the pail
and it bloody near broke the whole family's heart

you could go in the field and he'd always be there
just rubbing and licking your coat
and I'll freely admit when I came through the ring
I stood still for a while with a lump in m' throat

but that was last week and the kids soon forget
after all that's what farming's about
and to tell you the truth I feel better m'self
now that the cheque's in the current account . . .!

'. . . aye well they *did* say the showers would be isolated, Charlie . . .'

Just a Bad Dream

'Bought a brainless auld yow called Matilda
I dosed 'er injected and pilled 'er
every lambin' for years she would drive me to tears
'til a post 'tween the lugs went and killed 'er

    now as luck would have had it that day
    the local RSPCA
    came trundlin' past as the bitch coughed her last
    so what the hell could I say

the judge said you bad-tempered fool
are all peasants so heartless and cruel
just to kill the poor sod well surely t' god
you could use a more delicate tool

    Your Honour says I the deceased
    was an awkward cantankerous beast
    and even a bloke in a wig and a cloak
    would've belted the bugger at least

y' could tell that the jury were choked
when they heard how the old girl had croaked
but I made it quite plain I would do it again
'cos ye gods I was sorely provoked

then the judge called the trial to a halt
said "I'm sure that this crime's not your fault
during lambing I've heard 'tis the mind that's
    disturbed
– but six months for grievous assault. . . ."

'. . . now I wonder where the little buggers got this idea from Gladys . . .?'

That Woman

We're into the nineties and not much has changed
the divine Lady Margaret rides on
into the dawn of a new decade born
goes the woman who's never been wrong . . .

remarkable girl irresistible force
can you contemplate Denis's life
or imagine for fun how the story might run
if she'd been but a poor farmer's wife . . .

she'd be running the lambing with handbag on arm
reminding the shepherd and vet
there'd be no foreign breeds on her maiden seeds
nor anything moderately wet . . .

the contractor's combine would come to her first
while the neighbours would just wait dejected
and pity the grader who tried to persuade 'er
that half of her hoggs were rejected . . .

delivery wagons would turn up on time
the driver might help to unload
the slugs wouldn't dare to nibble wheat bare
and sheep wouldn't die right next to the road . . .

when money was tight and interest rates high
and the deficit way out of line
when the trade at the mart was breaking her heart
it's the chief auctioneer who'd resign . . .

if the plans didn't work if the statement glowed red
and the overdraft couldn't be met
well God help the firm who ruffled her perm
and declined to reschedule the debt . . .

she may walk like a yow with very bad feet
but she'd always be boss in the fold
and I'd even wager that old mother nature
would do what she's bloody well told . . .

she might go on for ever through tempest and drought
like some latter-day Queen Boadicea
with that smile on her face not a hair out of place
and her bum sticking out at the rear. . . .

'. . . do y' think it's maybe hardly dry enough . . .?'

'. . . my bird, I think . . .'

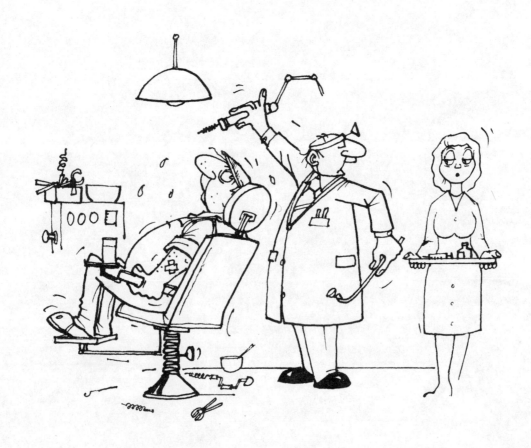

'. . . apparently he was biting some barley to see if it was dry enough for something called intervention, – and it seems it was . . .!'

The Good Life

For a life that is healthy
then farming's a must
provided of course
you keep out of the dust
that floats from the drier
and fills up the shed
stay clear of all that
and you'll never drop dead . . .

there's nothing like farming
fresh air every day
provided of course
you aren't caught in the spray
that drifts o'er the barley
when sudden breeze blows
right into the cab
and straight up your nose . . .

farmers are lucky
the job is a snip
provided of course
you don't swallow some dip
it can ruin digestion
as well as kill flies
and it isn't much fun
when it gets in your eyes . . .

for a long life the farmer's
a pretty safe bet
provided of course
you avoid paraquat
the wild suckler cow
a rampaging bull
the runaway tractor
the life's never dull . . .

farmers are colourful
and seldom grow paler
provided of course
they're not caught in the baler
worried and penniless
or comfort'bly wealthy
you can't tell for sure
'cos they always **look** healthy. . . .

'. . . well y' can't be all *that* wise if you've come this far without a reservation . . .'

Excuses Excuses

Christmas shoppin'? – you've got t' be jokin'
go yourself have some fun best of luck
just don't expect me t' come out on your spree
I'd rather stay here and shovel some muck . . .

Christmas shoppin'? – there's no way I'm comin'
and I won't change m' mind if y' nag
I would have t' get dressed put on a clean vest
and follow y' round with a carrier bag . . .

Christmas shoppin'? – y' cannot be serious
that Santa Claus bloke drives me mad
y' know bloody fine that he's no friend of mine
and he's cost me a fortune since I was a lad . . .

Christmas shoppin'? – it's only for townies
and women who just like to spend
yes I know Charlie goes but he's led by the nose
and it drives the poor fella right round the bend . . .

Christmas shoppin'? – we're not **made** of money
we haven't had **that** good a year
the bank's playin' hell and y' know very well
the heat in those shops always makes me feel queer . . .

Christmas shoppin'? – it's all aggravation
all those people I'll end up half-dead
just leave me alone I'm quite happy at home
I'll clean out the drains in the byre instead . . .

Christmas shoppin'? – who needs all these presents
your mother young Willie let's see
there's yourself of course pet and try not t' forget
a cap and a new pair of wellies for me . . .

Christmas shoppin'? – I just haven't time dear
I've got too many things to arrange
so I'll stay here at home off y' go on your own
. . . here's a fiver and bring back some change . . .!

'. . . and don't expect too much, – I hear Santa Claus had a disappointin' year . . .'

Just One More

You've caught and lambed the bitch at last
and mothered-up the pet
you pause a while and with a smile
   — you light another cigarette . . .

You've ploughed it slowly up and down
till all the furrows met
still miles from home and all alone
   — with only one more cigarette . . .

You're at the mart till half past three
and still bought nothing yet
with stone-cold feet and nowt t' eat
   — except of course a cigarette . . .

You're sitting sadly paying bills
accounts and cheques and debts
it's late at night and you just light
   — one more lousy cigarette . . .

When farming's not a lot of fun
to help you to forget
the soggy corn the blackest dawn
the calf stillborn
the wife forlorn
the trousers torn
the dog half-worn
the crumpled horn
the money gorn
   — y' need just one more cigarette. . . .

'. . . how fascinating, – I don't think I've ever met a two-year-old turkey . . .'

'. . . of course sometimes the new-born lamb is reluctant to suck, – and has to be . . . encouraged . . .'

'. . . well they're not goin' t' come anywhere near while you're still wearin' that bloody silly outfit, are they . . .?'

Life Gets Tedious

There's some mildew on m' barley
plague o' beetles on the swedes
pigeons eatin' clover
on m' patchy maiden seeds
the winter wheat's got eyespot
and the oilseed rape is thin
there's a dead yow in the meadow
couldn't catch her next of kin
the lambs are all on tip-toe
with their rotten scalded feet
a drain's blocked in the stackyard
and the collie dog's in heat
the garden's full of thistles
the lawn is two feet high
there's a bull calf coughin' gamely
but I think he's gonna die
the hens have all stopped layin'
and the bills keep comin' in
the co-op wants the telly back
and mother's on the gin
now I've got a runny nose
on top of all the rest
'cos I didn't wait till May was out
before I cast m' vest. . . .

'. . . come bye, y' stupid dog . . .!'

Bard to Verse

Not many farmers write poetry
proper peasants don't dash off a sonnet
there's no rhyme in his head when a sheep's lyin' dead
with a greedy old crow sittin' on it . . .

he isn't by nature romantic
very few care to break into verse
when the going gets hard they don't quote from the
   bard
much more likely a four-letter curse . . .

he'll maybe recall the odd lim'rick
or a snippet of prose learned at school
but a bloody great dollop of Shelley or Trollope
won't trip from his tongue as a rule . . .

he won't wander about like a Wordsworth
nor recite from the works of Molière
and an ode from the heart in the ring at the mart
well let's face it he just wouldn't dare . . .

when the combine is standing there knackered
and the bank's ringing up every day
there is damn little solace in Homer or Horace
and the overdraft certainly won't fade away . . .

yet I once knew a poor rhyming farmer
who wrote poems that nobody read
when he finally died on his stone they inscribed
a wee message and here's what it said . . .

"Here lies a dead peasant poet
who dreamed as he sat on this tractor
we're not sure why he's gone — but to plough the M1
must have been a contributing factor. . . ."

'. . . c'mon Sep, – it's no use, – she's a gonna . . .!'

Peasant's Law

When the corn is as high as a Charolais' eye
and the sunset's a promising red
when the forecast's set fair let the farmer beware
'cos there's bound to be mischief ahead
   ahead, there's bound to be mischief ahead . . .

when the combine's on song and just rollin' along
and the straw is all cracklin' and bright
there's an old peasant's law that says trouble's in store
and it's lurking there just out of sight
   out of sight, and it's lurking there just out of
      sight . . .

when the trade at the mart brings a song to your heart
and your sheep make a hundred a head
chances are that the sale has a sting in the tale
back home there's a yow lyin' dead
   quite dead, back home there's a yow lying dead . . .

when the overdraft's shrinkin' and you might be
      thinkin'
for once you've made more than you've spent
that's the day when the Duke makes a note in his book
and asks for a rise in the rent
   the rent, and asks for a rise in the rent . . .

when they're lambin' with ease not a sign of disease
and it's all gone accordin' to plan
well I can't tell you how but it's any time now
when the shit (pardon me) hits the fan . . .!

'. . . oh yes we've had a canny crop, – five solicitors, three accountants, a couple o' bank managers and a chief constable so far . . .'

'. . . now look what you've done, – it'll take **another** bloody miracle t' get them back . . .!'

'. . . invented it m'self, – most of them seem t' get the message after about two days of treatment . . .'

Any Fool can be a Farmer

Any fool can be a farmer
in fact it helps no end
to be a little crazy
or half-way round the bend
it's not essential to be crackers
but no need to be perturbed
as long as you're not normal
and obviously disturbed
if you can't meet these requirements
no need to break your heart
you'll quickly get the hang of things
once you get a start
if the weather can't destroy you
it'll be a calving cow
or a pompous politician
or an old demented sow
a collie dog that's clue-less
who doesn't know his name
or a sheep that's just plain awkward
they can all drive you insane . . .

but who could want a better life
of country skills and craft
so join our happy rural band
be destitute and daft
sell up all your assets
before it's far too late

you're just the bloke we're looking for
the perfect candidate
abandon all your stocks and shares
you must've seen the signs
any fool can be a farmer
— come this way I'll sell you mine . . . !

'. . . I think we might get a good run today, Colonel, don't you . . .?'

Happy New Year

What will it bring
what does it hold
will it be modest
extravagant bold
fruitful and friendly
famine or feast
a breeze from the West
a gale from the East
sunshine and showers
mountains to climb
a blizzard in spring
come next lambing time
how can we know
is it worth a wild guess
while we harvest in dust
or a damp clarty mess
haymake in heat-wave
bale muck in the rain
Wogan on telly
again and again
another Chernobyl
will battles be fought
will McEnroe argue
on packed centre court

will the oak beat the ash
will we beat Australia
will the year be a winner
or classed as a failure
 — of one thing I'm sure
it's a pretty safe bet
it disappears quicker
the older y' get . . .!

'. . . ups-a-daisy, Santa baby, – time to spread a little comfort 'n joy . . .'

Prayer for Today

Protect us dear Lord from all this hysteria
terminal paté and cheese with listeria
rife salmonella in all of our birds
and incomprehensible medical words . . .

   Save us now God from nuclear slaughter
   nationalised debt and privatised water
   and if you have time can you kindly repair
   that bloody great hole in the ozone layer . . .

Help us Almighty to clean up our rivers
give up the fags and take care of our livers
persuade politicians to be more discreet
lest everyone fears that there's nowt safe to eat . . .

   Preserve us our Saviour from lobster and crab
   poisonous prawns and cows that are mad
   keep our cholesterol reasonably low
   (and anything else you think I should know . . .)

Lead us O Father through all of this mess
confusion pollution and modern day stress
give us some wine and our daily brown bread
and just let us sleep in a warm comfy bed. . . .

'. . . gorn away . . .'